Books by Paul A. Johnsgard
published by the University of Nebraska Press

Birds of the Great Plains:
Breeding Species and Their Distribution

Birds of the Rocky Mountains

Crane Music: A Natural History
of American Cranes

Ducks, Geese, and Swans of the World

The Nature of Nebraska:
Ecology and Biodiversity

The Platte: Channels in Time

The Plovers, Sandpipers, and Snipes
of the World

Song of the North Wind:
A Story of the Snow Goose

This Fragile Land: A Natural History
of the Nebraska Sandhills

Those of the Gray Wind:
The Sandhill Cranes

Those of the Gray Wind: *The Sandhill Cranes*

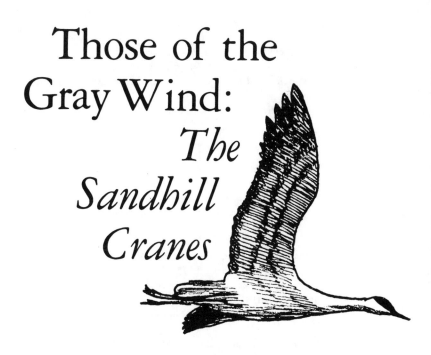

Paul A. Johnsgard

University of Nebraska Press
Lincoln and London

First Bison Book printing: 1986

Library of Congress Cataloging-in-Publication Data

Johnsgard, Paul A.
Those of the gray wind, the sandhill cranes.

"Bison book."
Reprint. Originally published: New York: St. Martin's
Press, 1981.
1. Sandhill crane. I. Title.
[QL696.G84J63 1986]   598.3'1   86-4292
ISBN 0-8032-7566-8 (pbk.)

Reprinted by arrangement with Paul A. Johnsgard

∞

For our unborn children:
may they too have cranes.

Those of the Gray Wind:
The Sandhill Cranes

Author's Note

Although the persons described here are fictitious, and any resemblance to actual persons living or dead is coincidental, the locations and descriptions of the places concerned are as factual as possible, as are the descriptions of the biology and behavior of the cranes. The folklore and legends are authentic, and all of the descriptions of crane behavior are based either on personal observation or on reliable descriptions by other persons.

Contents

 *Spring, 1860*

When our Earth Mother is replete with
living waters,
When spring comes,
The source of our flesh
All the different kinds of corn
We shall lay to rest in the ground
With their Earth Mother's living waters,
They will be made into new beings . . .
That our Earth Mother
May wear a fourfold green robe,
Full of moss,
Full of flowers,
Full of pollen,
That the land may be thus
I have made you into living beings.

Zuñi Prayer

North to the
Flat Waters

A bitter wind was driving out of the northwest, and occasionally a mixture of sleet and snow battered their wings, but the cranes pushed ever northward. They had left New Mexico Territory nearly two days before, and other than spending the previous night along the banks of the Arkansas River, they had been flying continuously. The flock was a gigantic one, numbering uncounted thousands of birds, all intent on reaching Nebraska

Territory and the Platte River, where their collective memory assured them of an abundant water supply and safety from coyotes and prairie wolves.

The young crane flew, as usual, between his parents, occasionally tilting his head to the side so that he could see the details of the land below. Once he saw a small herd of bison, and now and again a group of antelope would hear the cranes and briefly look up from their grazing to watch the birds pass overhead.

The land below was still partly covered by snow, especially on north slopes and in the ravines and valleys, where small creeks cut their way through the native prairies of the western Flint Hills. As the birds passed the Republican River, many of them pulled away from the main flock and circled the river several times, but there were no wide bars or islands on which they could roost. Thus, reluctantly, they struck out again into the eye of the wind, and set their minds toward the smudgy dark line that was the Platte Valley, which was somewhere just out of sight on the northern horizon.

The long, wavering gray line of cranes was like a giant aerial armada, weaving and advancing on Nebraska in wave after wave of birds, stretching as far as the eye could see—even beyond. Among them were cranes that had spent the winter months in flocks scattered from the Staked Plains of northern Texas and eastern New Mexico Territory southward into Mexico almost to the Central Valley of Mexico, and which were now all converging on a common goal, the broad and sheltering Platte. Here they would rest and feed in peace for more than a month, gathering energy for the second and much the longest part of their spring journey, which would carry some of them to the farthest coastlines of Canada and Alaska and even beyond, to the tundra of Siberia still thousands of miles away. They would need every bit of reserve energy for this tremendous journey, for food would become increasingly hard to find as they struggled against the last gasps of the Canadian winter in their push to reach the tundra just as soon as it was becoming free of its snowy prison.

The birds maintained a continuous din, as each family tried to maintain contact with all its members, and sometimes among the clarion *bir-rrrrrrt* calls of the adults one could hear the much higher-pitched notes of the yearling birds, whose voices even now were losing their baby quality and beginning to assume the rich timbre of the adults. There were differences among the adults, too. The largest birds, weighing up to ten pounds, had the deepest and the most penetrating voices. These cranes had the least distance to travel, only to the forests of central Canada, before they would stop to nest. But the vast majority of the birds, weighing seven to eight pounds, and with a wingspread of about six feet, were destined to fly the farthest, and these birds had the highest and most rattling voices. These were the "lesser" sandhill cranes or "little brown" cranes. Yet, they were hardly little, and none of them were yet brown. Not until they arrived at their nesting grounds would their plumages become stained with brown. Now they were almost uniformly ash-gray, save for the patches of bare red skin on the foreheads of the adults.

The young crane occasionally called to his parents, but mostly he remained quiet, saving all his energy for fighting the wind and cold. His long legs were freezing, and sometimes he would draw them forward up into his flank feathers to try to warm them, but flying this way was so awkward and the wind resistance so great that before long he would simply let them dangle behind him and suffer the cold.

As the afternoon wore on, the snow and sleet abated, and the sky gradually brightened to the west. As the visibility improved, the birds could actually see the Platte River, a silvery knife cutting gracefully through the prairies, lined with leafless cottonwoods and elms. The river was not so much a single stream as many separate channels. Each wandered eastwardly as if it had a mind of its own about the best way to reach its destination, sometimes joining another channel for a time, but often splitting away to follow an entirely different pathway. Thus, hundreds of islands were formed, some as small as a few acres, and others several miles in area. Many of the small islands were only barren deposits of silt

and sand, scoured free of vegetation by the
ice that had recently covered the channels;
others were just now becoming stabilized by
growths of low willows and other brush that
had gradually formed a protective barrier
against the yearly scouring action of the ice.
Still other islands were well grown to trees,
and from these thickets deer occasionally
bounded when they were being chased by the
few prairie wolves left in the valley.

     Since the earliest migrations west, the
Platte Valley had been a natural corridor for
mountain men, explorers, and finally,
immigrants. The Mormons had followed the
north side of the river on their way west from
Omaha to Utah Territory, and indeed a few
had been forced for various reasons to stop
and remain in the valley for a time before
gathering the strength and means to finish
their exodus to the Promised Land. And
along the southernmost channels of the river

there were already ruts being cut into the
prairie sod, where Conestoga wagons had
worked their way westward, usually starting
from Nebraska City, and frequently going as
far as their meager amounts of money and
equipment would carry them. In the 1850s a
number of German-speaking immigrants had
moved into the area from Iowa, and in 1857
they had formed a small community on the
north bank of the Platte. They had decided
to call it Grand Island, after the earlier fur
traders' name for it. Only two years later a
disastrous fire almost leveled the settlement,
which was later rebuilt farther from the river.
Slightly farther upstream, other Germans had
chosen one of the larger islands as a place to
begin homesteading, and had named it
Schumacher Island.

For the cranes, these few scattered sod
huts and the tiny herds of livestock were
insignificant. The settlers were much too

busy trying to survive to think about disturbing the birds. And the last bands of Pawnees were now being progressively forced out of the valley, particularly since the construction of Fort Kearny and the establishment of a permanent cavalry garrison to help protect the Immigrant Trail. For the cranes, as for the ducks and geese that had preceded them northward, the critical feature of the landscape was the Platte. Nowhere else in their entire spring migration would they be able to rest and forage in a river that had the combination of so many safe islands and bars for roosting, with so many wet, grassy meadows close to the channels where they could leisurely forage for germinating plants and seeds.

Here, too, the three-year-old birds, ready to mate for the first time, would strengthen their pair bonds with their new mates, by foraging together and sometimes dancing with them. Of course, dancing in cranes is almost as fundamental as eating, and even when the birds were only a few months old they had begun to bow and toss sticks or bits of grasses into the air whenever they became excited or uncertain of how to respond to a new situation. But now the

dancing would have yet another purpose, that of building a bond between the newly mated birds.

Soon the young crane and his parents were over the river, and could see that thousands of others had preceded them there. Many were still feeding in the fields and meadows nearby, while others had already returned to their evening roosting sites in the middle of the broadest and most shallow channels. The arriving birds began to call loudly, and were quickly answered by the roosting birds below, as if they were welcoming them and telling them that the area was safe. Perhaps because they were so tired, the arriving cranes spent little time circling the area, for it was slowly becoming dark and hard to see.

The vast numbers of arriving birds, and the tremendous din of so many thousands, all sounding nearly alike, was confusing, almost frightening to the young crane. As the birds began to descend and prepare to land, the young bird looked frantically about for his parents, but they had somehow become separated and were lost to view. The crane stopped his descent and flared back upward, calling wildly, but his tiny voice was drowned out by the great crane chorus surrounding him. Time and again, he circled the river, but gradually lost his bearings, and soon was not even certain whether he was in the right area. The entire channel below was virtually alive with cranes, all jostling for the best shoreline positions for spending the night.

As the western sky began to fade, nearly all the cranes were on their roosts, except for a few latecomers. For the adults that had visited the Platte many times, there were no great problems in landing. Following the deeper channels closely, they

drifted in against the wind and simply
gradually lost altitude and speed until they
could see a small bar or island on which to
land. But the young crane had never before
been forced to land in such a difficult place,
where the roosts were so close to tall trees,
and there were no broad and open approaches
to be followed. Finally, exhaustion and the
darkening skies made him decide to risk a
landing; later he could search for his parents.
Turning upstream into the wind, he would
make one last swing across the part of the
river where he thought his parents might
already be roosting.

As he worked his way westward, he
slowly lowered his altitude, balancing
caution against his eagerness to land and rest.
By now it was almost totally dark, the only
light that which was reflected off the river's
surface. Seeing a small bar with a few cranes
several hundred feet downstream, the young

bird tilted his right wing down and began to ease into a slow turn.

Intent on his chosen landing place, the crane could not see the old cottonwood snag, which a few years before had been hit by lightning and now stood gaunt and dead, with one large portion hanging out over the undercutting river. As the crane passed by, his left wing struck a branch, and he was thrown into a cartwheeling tailspin into the river.

Nearly knocked unconscious by the sudden blow, the crane was suddenly transformed from a beautiful flying machine into a drowning bird, struggling to keep afloat in the icy waters. The rapid current of one of the deeper channels beside the shore was pulling him downstream. Now his long wings were only a hindrance. His feathers were not waterproof and simply soaked up the water. Struggle as he might, the bird was unable to reach the shoreline or to attempt to fly. His right wing, although not broken, was so badly bruised that he could not extend it or pull it up into his flanks. Thus, the bird floated helplessly downstream in a sodden mass of flesh and feathers.

Gradually, the cold waters numbed the bird's strength, and he abandoned his struggles to escape, resigning himself to drowning. Then his legs touched bottom, and he found himself caught in a wire fence that had recently been strung across the river by some nearby homesteaders, who were enclosing a small piece of ground for their few cattle. Now at least he would not drown, for although the fence held the bird firmly in its grip, he was just barely able to pull his head and wings up out of the freezing stream. And thus the bird spent the night, half-unconscious, occasionally waking to struggle, then again lapsing into a near coma.

The next morning dawned bright and clear, but the young bird did not hear the cranes leaving their roosts at dawn. Instead, wet and cold, he lay huddled just above the water's surface, his head and neck over one strand of the wire, his wings caught firmly by the other.

In a sod-walled hut a few hundred yards away, Kristina Hahn was getting ready to do her morning chores. Twelve years old, she had accompanied her parents and brothers west just a year ago, and when they reached the tree-lined Platte it had reminded her immigrant parents so much of their original German homeland along the Rhine that they decided to go no farther. Kristina's father had told her she must check the fence, because just the week before the heavy ice flows on the Platte had nearly uprooted some of the posts, and there was danger that their few cattle might wander into the river.

Kristina put on her heavy jacket and, with a call to her dog, left the house. She heard the cranes calling in the far distance, but the birds were much too wary to have ever allowed her to approach them, so she scarcely gave them a second thought. Overhead, a skein of geese was making its way downstream, and their V-shaped pattern reminded her of an arrowhead she had found on the riverbank just a few days before.

Suddenly, her dog began an excited barking, just as she was reaching the river. It seemed to be directed toward a pile of gray debris washed up against the fence, perhaps a dead possum that had drowned in the water the night before. Coming closer, she saw that it had feathers rather than hair, and was apparently a dead crane. Yet the dog was acting slightly frightened of the object, which seemed to be weakly struggling to break free of the wires. Rushing ahead and wading into the water, Kristina reached down and lifted the soggy mess into her arms, carefully pulling away the strands of wire that had encircled its wings. Too weak to protest, the crane opened its eyes and through a semiconscious haze could feel itself being carried away.

With a cry of concern, Kristina said, "Oh, it's a wild crane, and its wing is all bruised and bleeding." With that, she gathered her jacket around the bird to hold it securely, and hurried back to the house just as rapidly as she could run, with her dog following closely behind.

Platte Valley Spring

The enveloping warmth of the girl's body and heavy coat brought the crane out of his deathly sleepiness, and he vaguely knew he was being brought into a closed place, and being covered by soft blankets beside a warm stove. He didn't try to raise his head, and he limply let his wings be extended as the girl's father examined him closely. The old man carefully tested each wing for possible broken bones, and ran his hand along the bird's body and legs.

"*Es ist ein wild Kranich*," he said, half to himself, for he was trying to avoid speaking German in the presence of his daughter. Turning to her, he said, "I think it has no broken bones, but its skin is torn and it has *ein wunde* on the arm."

"Can I keep it please, Papa? I promise to look after it and keep it with the chickens in the chicken coop, so it won't cause you any trouble. I'll catch grasshoppers for it and it won't eat any of the chickens' food."

The old man looked closely at his daughter. Her eyes were shining with excitement, and her face had a light that he hadn't seen since her brothers and mother had died of cholera the fall before, when she had barely escaped with her own life.

"Only as long as it is sick. When it is well, you must let it go, so that it can join the other *Kranichs*. It can only live happily in the wild, for that is the only life it has known."

"I promise, Papa." Kristina joyously lay her head down to touch the crane's soft back feathers, then quickly ran off to look for some ointment to put on its bruised wing and scratches.

When she returned, the crane was trying to stand up, but he was still too weak, and soon collapsed in an awkward jumble of legs and feet. Kristina couldn't help but laugh at his clumsiness, and decided that the bird should be placed in a tall wooden box that had once served for packing her clothes on the way west, and later had been turned sideways in order to serve as a stool. At the bottom of the box she arranged an old blanket to form a nestlike cup, and gently placed the bird inside.

"Now, I must try to find some food for it," she thought. She knew that there were no insects out yet, and decided that she would have to feed it some of the chickens' cracked grain after all. Running out to the small frame chicken coop, she filled a cup with the grain, and brought it back to the house, where she poured some of it out in front of the bird's bill. At first the crane did not seem to pay any attention, but as a small beetle that had been among the seeds began to scurry away, the crane looked down, made a tentative peck at the little black insect, and realized that the other objects were soft and edible. Slowly he began to eat, much to the girl's delight.

Kristina spent all of that day with the crane, except when her father called her away to help with the supper chores. There were still no schools in the area for her to attend, and so there was nothing to draw her away from this wonderful new pet. Her dog was both shy and jealous of the new creature,

which suddenly had stolen Kristina's
attention. The crane raised its bill
threateningly and gave a high-pitched broken
call whenever
the dog
approached
and tried
to be
friendly.
On the
other hand,
whenever
Kristina appeared
with food, the bird
would utter the same
plaintive food-begging
note that he used to make
when one of his parents
would find a morsel and
let him take it from its
bill.

After a few days, the bird was easily standing on his two legs, and spent much of his time exercising his sore wings. Often he would sleep for hours, sitting in his box with his legs folded under him, and with his breast and belly resting on the blanket below. His head was usually tucked into the long shoulder feathers, but at times he even slept with his neck extended and his head upright. As he awoke, he would slowly stand up, shake his plumage to ruffle and rearrange the feathers, and begin a long period of preening. Sometimes he would stretch one wing and the same leg while balancing precariously on the other leg. Or, he would stand for hours on a single leg, folding the other upward and hiding it from view in his long breast and belly feathers.

Kristina watched the bird with endless fascination. He was almost humanlike, greeting her always with excited rattling

notes and little bounding jumps as he tried
to leave the box to meet her. Soon he was
able to jump out of the box, and would run
quickly over the dirt floor of the room to
look for food in Kristina's clenched fist.
Sometimes, when in a hurry, he would also
spread his wings while running, and often he
almost ran into the wall on the opposite side
of the sod hut as he tried to stop. After all,
the entire building was only 24 feet long and
had a white sheet hung across the middle to
separate the sleeping quarters from the
kitchen and living room. The kitchen stove
provided the only heat for the house, but this
was more than enough, and during the
summer sometimes it was taken outside to
provide a little more living space in the tiny
dwelling.

One day, Kristina decided to bathe
the bird, to remove the last bits of dried
blood from his matted feathers. She brought

in a deep pail of warm water. At first the crane tried to take a drink, but soon stepped in the pail and tried to crouch down in it, moving his wings about and splashing water all over himself and the girl. After he was thoroughly wet, he stepped out in a dignified manner, and began to preen his feathers, starting near the tail and slowly moving forward. Kristina tried to help him dry his feathers with a small towel, and to her amazement he took the towel out of her hands with his beak, and began to rub it along his flanks and belly, drying himself.

Kristina laughed at this wonderful behavior, not realizing that the crane was simply doing much the same thing a wild crane does when it feather-paints itself with mud and grass in preparation for nesting. In this way, the bird's gray plumage is gradually stained to a reddish-brown from iron pigments in the mud, and the bird more effectively blends into its brownish spring landscape.

When Kristina told her father of the crane drying itself with the towel, he chuckled and told how, in the Old Country, there was a legend that when cranes slept at night in the water they always stood on one leg, clutching a stone in the other. Thus, if something should awaken one of the flock at night, it would automatically drop the stone, and the resulting splash would awaken and warn all the other birds of the flock. Kristina didn't believe this, but afterward she was amused to see the bird often standing and dozing close to the stove, sleeping soundly while standing on a single leg, with the other delicately balanced in the air as if it were holding an imaginary stone in its long toes.

Outside, the cranes made their regular morning and evening flights to and from the river, and as they passed over the house the young crane could hear their calls, and often answered them with his own much weaker call. Kristina listened each day with both pleasure and sadness, for she knew that soon she would have to let her bird go if it were to join the flocks on their flight north. It was nearly April, and the fields were starting to turn green. The cottonwoods and elms had begun to shimmer golden-green in the spring morning sunlight, and the whistling calls of the pintail ducks along the Platte were only a memory, for they were now pushing northward.

One day Kristina took the crane outside for a walk, and the dog began to playfully chase him until, half running and half flying, the bird skimmed above the grasses and found himself flying free. The crane looked downward in surprise; he had almost forgotten the delicious sense of being free of the ground and the feel of the flow of air around his wings. With a burst of power, he gained altitude, and began to circle back toward the girl and her dog, both running madly through the grasses below.

Kristina called to the crane, and the dog barked excitedly, but the crane could also see a gray line of flying birds in the distance. For a moment he hesitated, not knowing whether to fly back to the girl or whether to respond to the deeper urges he felt. But there could be no doubt; his world was the sky and the river, and his fate was to be free to find his family and to share his fate with them. With a quick look backward toward the girl, he called once, then struck out for the crane flock that was vanishing in the distance.

Kristina fell in the grass, tears streaming down her face, and the dog nuzzled against her and whined softly, not understanding what had happened. She hugged the dog in her arms and wept.

 *Summer, 1900*

And yonder, wherever the roads of the Rain
Makers come forth,
Torrents will rush forth,
Silt will rush forth,
Mountains will be washed out,
Logs will be washed down,
Yonder all the mossy mountains
Will drip with water,
From all the lakes
Will rise the cries of the children of the Rain
Makers,
In all the lakes
There will be joyous dancing . . .

Zuñi Prayer

Destination: Arctic

As the warming breezes of early April swept down the Platte Valley, the crane flocks grew increasingly restless. The great flocks of Canada geese and white-fronted geese had already left the valley. The river, fed by its headwaters in Wyoming and Colorado, had crested, and the spring surge of floodwaters had left muddy and gravelly deposits, where killdeers were building their nests. The soft music of courting mourning doves drifted down through the leafing trees.

The cranes were alert to these
changes, and with each passing day the flocks
had diminished noticeably. The birds were
also being ever more harassed by the settlers
along the river, for they were excellent spring
food, and there were still no effective laws to
protect them. The Grand Island settlement
to the north of the river was rapidly
becoming a thriving city. A few wooden
bridges had been constructed over the river,
and the cranes were finding themselves
progressively hemmed in by human
activities. Yet the river itself still ran nearly
bankfull, for there were no dams or irrigation
ditches yet to siphon away the life-giving
water and threaten the river's future.
Eventually, more than 40 dams would be
built along the Platte, and most of its water
would be diverted or otherwise lost. Year by
year, the dying river would become
increasingly shallow, confined to narrow

channels, and the muddy bars and islands would grow up to willows and cotton-woods.

None of the humans living along its shores had any inkling of the changes that were destined to occur, and indeed many of the farmers living there had already forgotten or perhaps had never even known of the dusky clouds of bison that once had forded the river in uncounted thousands as they wandered north with the spring, toward the lush and cool prairies of the Dakotas.

On a bright morning early in April, a vast flock of cranes arose from the Platte, caught a rising thermal, and began a long circling climb that carried them several thousand feet above the valley. From this height, the young crane could look downward and see for dozens of miles in all directions. To the east and west the Platte stretched to the very horizons, and to the

north the bird could see the gently
undulating hills that marked the edges of the
Nebraska Sandhills. They were still tinted
with a haze of gold and rusty brown,
produced by the last year's growth of
bluestem and other sand dune grasses that
lightly danced on the sand dunes, freezing
them firmly into place, like a painting of a
wave-tossed sea. To the south was the
Rainwater Basin, a marsh-flecked prairie so
flat that the spring meltwaters simply
remained caught in clay-bottomed
depressions rather than flowing away; in a
few decades it would be drained for farming
purposes.

As the birds struck out northward
over the sandy ocean of the Sandhills, the
young crane called to his parents from time
to time, and they answered back in kind.
His voice, now throaty and full, was nearly
that of an adult; the last few weeks had

produced a surprising change in its richness
as he approached his tenth month of life.
Yet, he still lacked the bright red crown
patch of the adults, and here and there rusty-
edged feathers on his wings and body gave
evidence that he was still less than a year old.
But during the weeks on the Platte, he had
gained much weight, and now was nearly as
heavy as his parents. They too had put on fat
during the six weeks in the valley, and in
spite of the blizzards that they were to face
during the remaining 2,500 miles of flight to
the Bering Coast, they should arrive in fine
condition for breeding.

As the flocks slowly worked their way
northward across South Dakota, they split
into progressively smaller groups. Some
began to drift eastward toward the Red River
Valley, heading for Hudson Bay and the high
arctic islands beyond. Others would cross the
center of North Dakota and strike out over

the vast boreal forest of the Canadian
interior, eventually to reach the ice-rimmed
shorelines of the Keewatin and Mackenzie.
But the young crane and his parents would
follow the valley of the Missouri River,
angling northwestwardly across North
Dakota and the prairie heartland of central
Canada. This route allowed the birds to
remain in open grassland areas for as long as
possible, avoiding both the snowy slopes of
the Rocky Mountains and the dark coniferous
forests of the Canadian lowlands, where
glaciers had scoured the land to its very core,
and where the rivers and lakes were held
rockbound and nearly devoid of edible plants.

Crossing the invisible border
separating the United States and Canada, the
crane flock continued on its way, intent on
reaching the grassy meadows of the South
Saskatchewan River before nightfall. There
they would rest for a week or two, to let

spring catch up with them, and again gather strength for the longest and most arduous leg of their spring journey. It was now late April, and the birds were little more than a week away from their destination. They would need only to wait for the right combination of winds and weather that would allow them to fly the last 1,500 miles along the crest of the Rockies until they entered the upper reaches of the Yukon Valley, and then follow it to the great tundra flats of western Alaska. Yet, many of the other birds in their flock would not be home even then. For them, a perilous flight across the Bering Straits, and a journey of a thousand miles or more across the frozen Siberian tundra were still ahead; they would be lucky indeed to reach their nesting grounds by the end of May.

The Tundra of
Igiak Bay

The coastal plain of western Alaska is more water than land; the flatlands between the deltas of the Yukon and the Kuskokwim rivers are an endless maze of lakes, streams, ponds, and marshes, with scarcely a hill to break the horizontal lines of this wild and watery landscape.

Yet, there are scattered ridges of upland tundra beyond the coastal flats. Even more rare are the few low mountain ranges that cast long shadows over the snowy lowlands in early spring as the sun begins its long voyage northward, and that hold snowbanks tightly clutched to their shaded northern slopes long into the warm summer months.

Such a mountain range is the Askinuks, which is born in the interior of Alaska nearly a hundred miles inland, grows as it sprawls northeasterly, and finally throws itself headlong into the Bering Sea at Cape Romanzof. Below its sparsely wooded slopes are the vast lowland tundras of Igiak and Hooper bays, one of the most isolated and pristine portions of western Alaska. Here there were no abundant animals that might attract fur traders, too few whales in the shallow coastal waters to support whaling vessels, and there was no gold in the scattered mountains to bring in prospectors.

Thus, the land lay primeval,

untouched by white men, even as the
Eskimos much farther north were becoming
influenced by prospectors, traders, and
whalers. And thus the tundra remained a
paradise for breeding birds during the short
but spectacularly beautiful summers.

The pair of cranes had been part of a
flock that arrived at Igiak Bay in early May,
when the first arrivals trumpeted their
approach to a deaf landscape of snow and ice.
The flock soon broke up into pairs, and each
pair returned to the territory it had occupied
the year before. Snow covered most of their
nesting areas, and each day the birds flew to
the sunny south-facing slopes of the Askinuks
to forage for the few shoots of green
vegetation that were tentatively poking their
tips above the lichens and dead grasses
covering the hillsides.

One pair, breeding for the first time,
staked out their claim to a piece of tundra

that adjoined the small river meandering its way through the flats behind Hooper Bay. Standing on a small knoll, the male trumpeted his proclamation of ownership. He was quickly joined by his mate in their unison call. The female's call consisted of two notes that were shorter, higher-pitched, than the more drawn-out calls of the male. As the birds harmonized, the male reached down, grasped a bit of dead vegetation, and threw it high in the air. Then, in a stiff-legged gait he bounded high in the air, spreading his wings to cushion the return, all the time facing his mate. She did the same, and together the two birds performed an exuberant dancing and calling duet that echoed away to the edges of the coastline. Other cranes in the area responded with similar calls.

As the days of May passed, the sun rose ever higher, and remained ever longer above the horizon, while the snowline on the south face of the Askinuks inched slowly upward. The daily snow changed to daily rain, and soon the small creeks had broken their icy bondage and were running bankfull, carrying the silt, branches, and other debris that had been accumulating since freeze-up the previous September.

By the last week of May the lowlands were all free of snow, and the ice on the ponds was disappearing, although they still froze over sometimes at night. Now the ducks, who

had been waiting patiently for the thaw in
the open leads between the ice flows along
the coastline, began to flood inland. Among
the first were the old-squaws, called A-hung-
ree-yak by the Eskimos, for the yodeling calls
of the long-tailed males. They are as much a
part of the tundra scene as the cranes. There
were also the Kow-uks, or spectacled eiders,
the males resplendent in their golden-green
crowns and their curious white spectacles,
while the drab females had the brownish
tones of dead grass and only a faint imitation
of the males' head pattern. Here too were the
arctic geese, including Not-ch-flick, the
emperor goose, Nuck-la-nock, the brant
goose, Too-tin-eye-yuk, the tiny cackling
goose, and Nuck-luck, the white-fronted
goose. All these birds gradually shifted from
the coast inland, the brants scarcely moving
beyond high-tide line before searching out
nest sites, while the emperor and white-
fronted geese moved into the freshwater

marshes, each to its own favored nesting area.

Above the tundra the shorebirds began their spring territorial display flights. Like tiny balloons, the pectoral sandpipers floated about in the sky, uttering the low *Doom-doom-tag* calls that gave them their Eskimo name. The black turnstones, or Chilee-muck, resembled giant black butterflies as they flitted above the tundra. By comparison, the tiny western sandpipers, the Ee-u-ga-guk, were like mosquitoes, uttering gentle musical notes and buzzy trills as they patrolled their own little world. Largest of all were the beautiful godwits, the T'wgo-ti-woo-guk, whose loud voices warned all other birds of possible danger, and whose rich rusty-brown plumage glistened like burnished copper as the birds made their territorial flights above the higher ridges of the upland tundra.

None of these birds were of any real

concern to the pair of cranes, who set about
building a nest as soon as the first parts of
their territory became snow-free. It was a
simple structure, built by piling dead grasses
into a shapeless heap beside the edge of a
small pond. Both birds added to it until the
nest was nearly four feet wide and almost a
foot high, resembling a soggy haystack. The
female stood on this pile of vegetation and
trampled the middle of it until a shallow cup
was formed; this would allow her to settle
down on it and, when she was crouched as
low as possible, her brown-stained body
would seem simply an extension of the dead
reeds below her.

The nest was hardly finished before
the female laid the first of her eggs on the
last day of May; two days later she laid her
second and last egg. The eggs were large and
handsome: greenish-brown in color, with a
variety of darker brown spots, freckles, and
blotches in random patterns. Unlike the

ducks and geese that were now nesting in the same area, the female began to incubate as soon as the first egg was laid. She did not line her nest with down; instead, the eggs lay side by side beneath her, her breast feathers cushioning them from the weight of her body, which gently touched them and provided them with life-giving warmth.

As soon as she began her egg-laying, the male paid close attention to his mate and with the laying of the second egg, he quietly approached the nest and gently eased her off it. After that the pair took regular turns incubating the eggs, usually in shifts of several hours, but with the female spending the night on the nest. The nest was never left untended, for the gulls and other predators were constantly in the sky, searching for untended nests that could be robbed to provide food for themselves and their own young that would also be hatching soon.

This was much the most dangerous

time of the year for the cranes. Bound to
their nest and territory to protect their eggs,
the birds took turns standing guard whenever
they were not actually incubating; often the
guarding bird would stand as far away from
the nest as it possibly could, while still being
able to watch for any approaches by foxes or
humans.

The nest was several miles away from
the nearest human habitation, a group of
small huts huddled at the base of the
Askinuk Mountains near Cape Romanzof,
and was also protected by several small
streams from any wandering dogs that lived
in the village of Hooper Bay. But every day
Eskimos would walk or kayak out into the
tundra marshes to look for the eggs of geese,
swan, shorebirds, and cranes. Crane eggs
were considered by the Eskimos the greatest
of all delicacies, even finer than those of Ko-
ute, the majestic white swan, which only

rarely attempted to nest in the area.

Each day a young Eskimo from the huts at the base of the mountains would kayak up the winding creek to look for eggs to gather for his family, and each day he would return with a few dozen duck, goose, and shorebird eggs. Luckily for the cranes, his visits had not yet reached their territory, and from several hundred yards away the guarding crane would watch the

approaching kayak and utter a soft warning
note to its mate. With that, the sitting bird
would huddle down even closer to the nest,
and lower its head so that it was touching
the side of the nest. Thus it would remain,
as motionless as a rock, until its mate uttered
an "all-clear" signal.

The long spring days passed into one
another almost imperceptibly. The sun
swung about in a great circle in the sky,
reaching its highest point at midday, when it
hung directly over the Askinuk Mountains,
and dipping down in the late evening nearly
to disappear behind the snow-clad peaks of
Nunivak Island, nearly 100 miles to the
south at the edge of the horizon. As hatching
approached, the female stayed ever more
strongly on her nest, and finally at the end of
June, just a month after she had laid them,
the eggs began to hatch.

The first laid was also the first to
hatch, and soon, as the chick pecked from

within, a tiny series of cracks began to
appear on the larger end of the egg. The
cracks soon encircled the upper part of the
egg. As the long task neared completion, the
chick pushed the top of its head against the
inner surface of the shell, gradually forcing
the top open. Then, with a mighty heave, it
kicked away the main part of the shell, and
soon lay free in the nest. It was wet, cold,
and panting, but was breathing air with its
own lungs for the first time. Tenderly the
female stood up, picked up the shell
fragments and crushed them with her beak,
swallowing most of them herself. The rest
she left, to break into tiny pieces and feed to
the chick later.

While the chick lay in the nest, its
down slowly began to dry, and as it did the
baby was transformed from a wet and
shapeless object into an enchanting ball of
golden fluff, with enormous pink legs, a
blunt yellowish bill that still had its white

egg-tooth attached to the tip, and large brown eyes surrounded by pale spectacles. It slowly sat up on its haunches, but found that the weight of its mother prevented it from rising any farther. Gradually the chick worked its way to the side of the nest, and poked its head out from the flank feathers of its mother. Getting a first look at the world, it could see little but grasses and water. As the chick's eyes began to focus clearly, it could see its father, standing some distance away and closely watching the nest. As soon as the eggs had begun to hatch, the female had become noticeably more nervous and reluctant to stand, but now with her first-hatched youngster moving about underneath her she finally stood up and gazed down on her offspring and the second egg, which was also starting to crack open.

The chick looked up toward its mother, uttered a faint peeping sound, and then moved back between her legs, wanting to be warmed again. Thus the mother settled back on the nest, and soon the baby chick climbed into the warm pocket between her folded wing and her flank. From this cozy spot the baby could sleep comfortably, and at the same time could easily extend its head out from between its mother's feathers to see the world around it.

During the next several hours the female sat quietly while the second egg hatched, and by the following morning both chicks were fluffy and dry. By then the older one, a male, had already taken a short trip from the nest, to swim to his father, who had spent the evening hours only a few yards away from the nest. A few hours later the younger chick, a female, was also clambering about on the back of her mother. Soon it became apparent that the nest was simply not large enough for raising an active family of two young cranes.

The young cranes had hatched none

too soon, for the very next day the young
Eskimo boy, who had been nicknamed
Eneepuk, or Snowy Owl, because of his sharp
and piercing eyes, came kayaking up the
creek in search of eggs.

The mother was still on the nest with
her two chicks when her mate uttered his
first alarm call. Paddling quickly up the
narrow channel. Eneepuk heard the male's
call, and knew that it signaled a pair of
cranes in the close vicinity. Then he saw the
brownish heap at the edge of the pond, and
his heart soared with delight, for he knew
that he might be able to take home some
prized crane eggs for his mother. Pulling the

kayak to land, he was about to leave it when
suddenly the mother crane left her nest,
exposing the two baby birds. Disappointed
that there would be no eggs, Eneepuk simply
stayed beside the kayak while the large bird
staggered away with its wings dragging
almost as if they were broken. She headed
toward the male, who was quickly
approaching to help defend the chicks. These
were now crouching in the shoreline
vegetation, unsure of whether they should
run or hide.

As the female crane reached her mate,
she stopped and, strangely, began to dance
before him, hopping and jumping about with
her wings still drooping. The male responded
in the same way, and together the two birds
performed their odd dance as the boy
watched in amazement. Then he pulled out
an arrow from its birdskin quiver and
wondered if he should try to shoot one of the
dancing birds. But his arrows were designed

for smaller birds, and the more he watched
the cranes the more he was touched by their
curious waltz. Almost without thinking he
began to beat the arrow on the sealskin deck
of his kayak, and to chant a dreamy song
that his mother had taught him about cranes.
As he beat his improvised drum, the birds
seemed to respond to its persistent beat, and
as he sang faster the birds speeded up their
dance; whenever he slowed down the birds
did too. Finally, the boy's curiosity was
satisfied, and he reluctantly got into his
kayak to return home with the eggs he had
already obtained.

When he reached home, he ran up to
the sod-covered hut and entered the darkened
room where his mother and older sisters
lived. His father and older brother lived
separately from the women and children in a
clubhouse, or kashim, where they slept and
made arrows, kayaks, and other important
items, while the women and children

provided them with food. Entering his hut through the low wooden doorway, Eneepuk scarcely noticed the overwhelming smell of seal oil that permeated the room. The only light came from a nearly transparent skin-covered hole in the roof; otherwise, the ceiling was made of logs and driftwood gathered from the beach, covered over with sod to provide a leaky shelter. His mother was sitting on a wooden bench, sewing a new parka with sinews from a seal, and using the soft belly skin of a goose from which she had plucked the larger feathers, exposing the luxuriant down underneath.

"Mother, Mother, I have just made the cranes dance," exclaimed the boy. "I made them dance both fast and slow, and they kept time to my rhythm as I beat it out."

The woman looked up and smiled. "You should spend more time looking for eggs, and less time doing such foolish things as making the Co-chee-sluck dance," she said.

"But Mother, they danced just like humans," Eneepuk replied, settling at her feet.

"That is because they once were humans," said his mother. "They are the souls of persons who once were humans, and will be humans again in some future life, but for now they are cranes. Yet, they remember that they once were human, and sometimes wish that they were so still."

The woman paused to collect her thoughts. Then she continued, "Once, long, long ago, when the cranes were getting ready to leave this land and fly south for the winter, they saw a beautiful woman standing near our village. They thought she was very desirable, and all the cranes gathered about her, lifting her into the sky with their strong wings. Some of the cranes flew in tight circles below her so that she would not fall or be seen, and they all screamed loudly, so that nobody could hear her calls for help. Finally, she disappeared with them in the sky, and was never seen again. That is why the cranes always fly in circles and call loudly when they leave here each autumn, remembering the time that they carried away our beautiful daughter."

Eneepuk listened, but his mind kept returning to the marvelous sight of cranes dancing in the tundra, and he wondered what they might be doing now.

In fact, the cranes had abandoned their nest site. The single experience with the Eskimo boy was enough to make them begin a long walk across the tundra, the male in the lead, the two chicks close behind, and the mother in the rear. Remaining close enough to their parents to prevent any danger from gull attacks, the chicks often peeped in distress as their tiny legs could scarcely keep up, even with the slow pace their parents set. Furthermore, they were always hungry, and could not resist stopping to peck at a midge or mosquito whenever one came into view.

At last the adults were satisfied that they had moved far enough, and the family settled into its new surroundings. Each of the parents took charge of one of the chicks, for the older male was inclined to bully his younger sister and to fight with her over food. Because he was nearly a day older, he was both larger and stronger, although both birds were growing rapidly.

By the time they were a month old, the two chicks had increased in weight from about five ounces to nearly three pounds, and they stood nearly two feet tall. They could run quickly through the tundra, and were safe from nearly all predators except, perhaps, foxes. A few weeks later their wing feathers began to appear, and by the time they were two months old the birds were attempting their first short flights. By then they stood nearly as tall as their parents, and weighed more than five pounds. The golden down colors had faded, and were being replaced by rusty-tipped grayish feathers, much like those of their parents. Their crowns, however, were entirely covered with short feathers, and their legs were gradually changing from their original pinkish tones to a dull gray.

Each day, whenever their parents called them, the chicks would reply with the baby peeping calls, but as the birds grew, these calls became stronger and began to acquire a more rattling character, sounding like a prolonged *pe-e-e-e-rrr.*

As July ended, the chicks were able to take prolonged flights with their parents, and they often flew to the upper slopes of the Askinuk Mountains, where they foraged in the lush tundra with other crane families. They spent more and more time with these other cranes, for they too had abandoned their territories and were trying to spend as much time as possible in foraging and putting on weight in preparation for the long flight southward. All of the adults had completed molting their body and wing feathers, and soon the newly grown ones would be fully regrown and ready for the long fall migration. By early August, the tundra was surprisingly quiet. The male oldsquaws, eiders, and most of the other sea ducks had long since gone back to sea, where they would spend the next ten months. Geese of many species cautiously convoyed their half-grown broods through the overgrown marshes and creeks, as they too made their way toward the coastline. The godwits, sandpipers and turnstones vanished into heavy cover as soon as their chicks had hatched, and would soon be winging toward shorelines as much as half a world away. The grouse-like ptarmigans, nearly snow white only a few months before, had changed into a tundra-matching fall coat of grays and browns that rendered them almost invisible to the sharp eyes of snowy owls and hawks, even as the sun began its slow curve back toward the south, and the days began to grow perceptibly shorter. For nearly all the birds, the summer was over, and a new generation of their kind had to be escorted to warmer lands. Even as the last remnant patches of snow were melting on the sheltered north slopes of the Askinuks, the arctic summer faded to only a memory.

 *Fall, 1940*

That yonder in all our water-filled fields
The source of our flesh,
All the different kinds of corn
May stand up all about,
That, nourishing themselves with fresh
water,
Clasping their children in their arms
They may rear their young
So that we may bring them into our houses;
Thinking of them toward whom our
thoughts bend—
Desiring this,
I send you forth with prayers . . .
From wherever my children have built their
shelters,
May their roads come in safety,
May the forests
And the brush
Stretch out their water-filled arms

And shield their hearts;
May their roads come in safety
May their roads be fulfilled.

Zuñi Prayer

The Roof of the Continent

Autumn comes rapidly in Alaska. Even as the cotton grass is still blooming, and as the last-hatched ducklings and goslings are practicing their first flights, the weather turns deathly cold, with the inky clouds that perpetually hang over the Bering Sea like a smoky blanket spread inward to the land, to shed their heavy loads of snow on a silently protesting landscape.

For the cranes, the first flights of the

young in late August came none too soon. A blizzard was developing over the Bering Sea, and the wind was increasing out of the northwest; it was time to leave. The two adults called to their gangling young one morning in early September, and suddenly they were on their way, together with the rest of the cranes that had summered near Igiak Bay. All told, they made up a flock of several dozen birds, which circled ever higher and higher, until they were well above even the highest crests of the Askinuks, before striking out in an easterly direction. They were flying with the wind, which made it easier for the young birds to maintain the brisk pace set by the adults. Before long they had breasted the Kuskokwim Range, and had bypassed the highest parts of the Alaska Range to the south, where Denali, the greatest peak of them all, emerged from its sheltering clouds like a white-robed ghost.

On and on the birds flew that first
day, past the snowy Wrangell Mountains,
and finally stopped to rest for a few days by a
mountain-rimmed lake in southern Yukon
Territory. The young birds' wings were
aching, for they had never before flown so far
or so hard before, and they were frightened
by the swirling winds that tossed them about
like leaves as they crested the highest
mountains.

This was strange and wild country,
quite different from the flat tundra to which
they were accustomed, and the cold waters of
the lakeshore where they roosted had few
plants on which they could feed. They would
thus stay only long enough for the youngsters
to rest, and early the next day the birds were
on their way again.

With each valley they crossed they
encountered more cranes, which often joined
their noisy flock. Soon the birds numbered

several hundred strong as they winged their way southeastwardly across the peaks and forests of northern British Columbia.

By the end of their second day of flight, the birds emerged from the mountains into an enormous plain, punctuated and rimmed with aspens that were turning golden in the crisp fall nights. The flock dropped down to a small tributary of the Peace River, where they joined the flocks of waterfowl that were concentrating for their own fall flights southward. Among them were trumpeter swans, whose young were only now gaining the power of flight, and would also be heading for warmer lands soon.

The birds were finally less than a day's flight from their first real opportunity to rest and forage. This destination was the upper reaches of the South Saskatchewan River, at the northern limits of true prairie. Here, long fingers of the prairies stretched northward to meet the aspens in a silent battle for dominance. In some places the aspens had won, slowly covering and shading out the prairie plants. In other areas the prairie firmly held sway, choking out the aspen seedlings before they could become high enough to cast their own shadows on the grasses. Thus, century after century the silent battle had continued, the aspens winning an occasional round in the wetter years, the grasses gaining ground whenever a prairie fire raced across the landscape and blackened the aspen forests, or when drought killed the delicate aspen seedlings before they could become established.

Now too the birds were starting to see signs of human activity. Most of the prairie areas were no longer actually native grassland, but instead were pastures, fallow lands, and grainfields. Here the land and farmers were slowly recovering from the long years of drought, and the first good crops of wheat in more than a decade were being harvested.

For the cranes, the newly harvested wheat fields were a bonanza. The threshing machines had left great areas of stubble that were low enough for the birds to walk through easily without limiting their vision, and waste grain was scattered abundantly on the ground. The birds gorged themselves, not only on the grain, but also on grasshoppers and beetles that they often found while searching among the stubble.

Each morning for nearly two weeks the flock remained in the area, spending the nights along the shoreline of a shallow, alkaline lake and flying out each dawn to the wheat fields. After a few hours they would fly to the lake to drink and pass the hottest part of the day, then in midafternoon go back to the fields a second time, to feed until nearly sunset. The birds gradually regained the energy supplies that they had spent in crossing the mountains of Alaska and northern Canada, and were slowly building up new fat reserves that would carry them to their wintering grounds still more than 1,000 miles to the south. But they had already come more than halfway: the cold

tundra of western Alaska lay almost 2,000 miles behind them, and there were no more high mountains to cross. For the rest of their flight they would be in prairie country, rich in marshes and wheat fields. And the cool September days of the northern plains did not carry the threat of unexpected blizzards and bitter north winds.

Among the flocks of cranes that continued to increase daily in their numbers were a few whooping cranes. These larger, white birds with black wing-tips were the last pitiful remnants of flocks that once were nearly as numerous as the sandhills, but had been persecuted to the very edge of extinction. Now barely two dozen birds made the annual trip from their wilderness spruce

and muskeg nesting areas of northern Alberta to southern Texas. For a short time, they mingled here with the sandhill cranes, lording over them and brusquely edging them aside whenever the two species were close together, for they stood nearly a foot taller than the sandhills. Only a few of the whooper pairs were leading rusty-toned youngsters, for the flocks had been so badly ravaged by hunters that very few breeding pairs were still intact. Most of the birds were old, non-breeding, adults that had lost their mates to some misfortune, and had not attempted to pair again.

Compared with the rattling calls of the sandhills, the great bugling trumpets of the whoopers that rang out each morning as the birds prepared to leave for feeding, or when they saw others of their kind in the sky were imposing indeed. The young sandhills often trembled as they heard the whooping cranes' piercing calls, and moved still closer to their watchful parents. Like the sandhills, the whooping cranes had once nested all the way from these portions of western Saskatchewan south and eastward across southern Manitoba, North Dakota, Minnesota, and Iowa to northern Illinois. But even before the sandhills had surrendered to marsh drainage, egg collecting, hunting, and finally poaching, the whooper flocks had retreated to the north. The final survivors had chosen to make their last stand in the wilderness of Wood Buffalo Park, where their nesting areas still lay undiscovered by humans.

As the fall days shortened, the nights became longer and colder, and nearly every night, frost would form and ice begin to develop around the edges of the cranes' roosting ponds. Thus, in early October, the two adult cranes and their youngsters joined a flock that headed south across the North Dakota–Saskatchewan border, and made their way over the gently rolling prairies of the Souris River valley. They were headed for the shallow alkali lakes of Kidder County, where an old glacial outwash plain provided a mixture of white alkali sand flats, shallow ponds, and surrounding prairie grasses. There avocets, godwits, and phalaropes had flocked during early summer to breed, and there too a variety of ducks and other water birds fed in the salty ponds among the few plants that could tolerate these alkaline waters. But for the cranes, it provided open vistas that gave them safety, and plenty of food could be found in the nearby wet meadows.

Rendezvous at Horsehead Lake

The teenage boy stood shivering in the crude reed-covered blind, almost knee deep in water. His left hip boot had a slight leak, and water had seeped in almost to ankle level. All day he had been feeling miserable. Not only was he cold and wet, but it was his very first duck hunt, and he was desperately trying to live up to his father's expectations of him. His dad had given him a new 20-gauge double-barrel shotgun for his fifteenth birthday, and ever since, the opening day of duck season had been circled on his father's calendar.

"Hunting will make a man of you," his father had told him. After all, his father was president of the local Ducks Unlimited chapter at Bismarck. Also, he owned the local Nash car dealership, and it was good business to talk about duck hunting with his customers.

The day had been almost a complete disaster. Jerry had started out with a boxful of shotgun shells in his ammunition belt, and now he was down to only a few rounds, including some special extra-powerful buckshot loads that his father had told him to save for geese, in the unlikely event that any had yet come down from Canada.

They were hunting in Kidder County, north of Dawson, and about 50 miles east of Bismarck. They had gotten up at 3:00 A.M., in order to be in the marsh well before dawn. The day's legal shooting had begun a half hour before sunrise, when the ducks were still only shadowy figures in the sky, but made relatively easy targets while they circled about in confusion in the half-darkness, trying to find a safe place to land.

Jerry's father, an experienced wing-shot, had already killed his limit of ducks and, rather than embarrass the boy by then shooting his birds as well, had ambled off toward a thicket of heavy cover about half a mile away, where he had seen a pheasant fly into the brush.

Jerry thought back on the day. So many times he had missed easy shots, and so often he had scared the birds by not keeping his head low enough, invisible, as they were approaching the decoys. He resolved that before the day was over he would have at least one mallard to take home. Every last shot would be taken with special care. He fondled the glistening stock of the shotgun, and imagined himself looking down the rib between its barrels at a gigantic goose.

By then his father was out of sight, but soon he heard three quick shots ring out, as the 12-gauge automatic shotgun fired in rapid succession. By the time the boy had turned to look, the cock pheasant had almost struck the ground, and a few feathers were still floating in the air to mark the spot where it had been aloft.

He had hardly turned back toward the decoys when a new sound reached his ears, and from the north a thin, wavering line of birds was approaching the lake. "Geese," he said to himself excitedly, and he quickly crouched down so that the water was nearly coming in over the top of his boots. He didn't dare look up, but the sounds of their voices grew louder and louder; they must have seen the decoys by now, he thought. Keeping his body as still as possible, and his head lowered to keep his face out of view, Jerry began to breathe quickly; he felt his heart pounding so wildly that he was certain that the birds would hear him and flare away.

But their voices kept getting louder, so that it seemed they must almost be on top of him, and he began shaking so much he could scarcely stand. Slowly he reached into his pocket for the two special buckshot loads that his father had given him, and slipped them into his shotgun to replace the ones in the gun. Then, as the cries of the birds reached a fever pitch above him, he stood up and threw the shotgun to his shoulder. Above him was a wildly flaring group of great gray-bodied birds, beating their wings in confusion and fear. He brought his shotgun to bear on one of the largest in view,

but as usual he didn't remember to aim in front of it, and with a convulsive jerk he closed his eyes and pulled both triggers simultaneously.

The young female crane was flying only a few feet behind her father, and as the boy stood up in the marsh she wasn't even aware of the danger. Suddenly both charges of the buckshot struck her body, ripping through her liver, heart, and abdomen, and shattering both of her wings. The massive blow killed her almost instantly, and she was quickly transformed into a shapeless mass of feathers as she collapsed and began to tumble toward the water below.

The combined recoil of both barrels had thrown the boy off balance. As he was struggling to avoid falling completely into the water, he could see the crane crash into the middle of the group of decoys. With a whoop of excitement, the boy waded out into the water, and grabbed the dead bird by its long neck. In his excitement he scarcely noticed the long legs and feet, or the bird's sharply pointed bill. While he was bringing the bird back to shore he saw his father walking up the shoreline, a pheasant swinging from the game carrier on his belt.

"Hey, Dad, look at the big goose I've just shot," Jerry yelled.

His father, now nearly back to the blind, looked at the boy and the bird, and immediately noticed the long legs trailing in the water.

"My God, Jerry, what have you shot? That's not a goose, it's a damn shitepoke or something!"

Jerry looked down at the bird for the first time, and noticed the blood dripping off the end of the pointed beak, which certainly didn't look like the bill of a goose.

"What's a shitepoke, Dad?"

"It's a kind of a heron that eats fish and such, and worse than that, it's against the law to kill them."

Jerry felt his world collapsing. Rather than finally pleasing his father, he had only disappointed him again. The bird he was carrying suddenly felt like it weighed a ton.

"Can I at least take it home and have it stuffed for the state museum?"

"For God's sakes, no; if a game warden saw us with that bird he would take away our guns, our hunting licenses, and everything else. We would be lucky to get away with a stiff fine. Get rid of that bird just as fast as you can, before somebody sees us with it. Throw it in a pile of weeds, and we'll head for home."

Jerry looked helplessly about the bleak shoreline, and decided that the clump of dead branches and weeds that had served as his blind was as good a place as any to leave it. He placed the bird gently down, with its head above water and resting on its bloody breast feathers, wiping away a clot of blood that was forming at the end of the bill. As he adjusted the bird's matted feathers, he said softly, almost to himself, "I'm sorry, I'm sorry, I'm sorry. I'll never ever do anything like this again."

The two walked back to the car in silence, and as Jerry broke open the gun before getting in the car he looked at the two spent shells that were still in the chamber. He removed them slowly and stared at them in disbelief, as if it were impossible that such small objects could have caused so much destruction. The smell of burned gunpowder clung to the shells, and now instead of thinking that it was a strong and manly odor, Jerry felt only nausea.

He climbed into the front seat of the car beside his father, who had just thrown the decoys into the trunk. As the car roared to life and slowly started forward, Jerry could stand it no longer. Hiding his head in his arms, he turned his face to the backrest and sobbed quietly.

The crane flock, thrown into sudden panic by the shooting, had scattered in all directions. The pair that had lost one of its two offspring was completely confused. Both the adult cranes and the remaining youngster repeatedly flew back above the marshland, looking hard for some sign of their missing member. The fatal shooting had been so very unexpected, and the female's death so sudden, that they failed to comprehend what had happened. When they had heard shooting earlier, as in Alberta and Saskatchewan, it had always been in the distance and directed toward waterfowl, and the birds were less wary than they might otherwise have been.

Time and again, the parents flew over the marsh, calling for their missing offspring, but there was no answering call. Below, the female's body was stiffening in the cooling evening, and flies were gathering around the gaping wounds. That evening a red fox caught the crane's scent, pulled the bird to shore, and dragged it to its den to feed to its family. In the marsh, the only remaining traces of the crane were a few rusty-edged feathers floating on the water and resting lightly among the shoreline weeds.

 *Winter, 1980*

That our Earth Mother may wrap herself
In a fourfold robe of white meal;
That she may be covered with frost flowers;
That yonder on all the mossy mountains,
The forest may huddle together with the
cold;
That their arms may be broken by the snow,
In order that the land may be thus
I have made my prayer sticks into living
things . . .
Following wherever the roads of the Rain
Makers come out,
May the ice blanket spread out
May the ice blanket spread out
May the ice blanket cover the country;
All over the land
May the flesh of our Earth Mother
Crack open from the cold . . .
Do not despise the breath of your fathers,
But draw it into your body.

That our roads may reach to where the life-
giving road of our Sun Father comes out
That, clasping one another tight,
Holding one another fast,
We may finish our roads together . . .

Zuñi Prayer

The Valley of the Sacred River

The afternoon was like most others in early November. A cooling wind came briskly out of the east, the gentle slopes on the hillsides were becoming muted, as the desert grasses turned brown and their colors were obscured by the silvery green of the ever present sagebrush. Above, the blue sky was strewn here and there with the wispy cloud manes and tails of wild stallions. To the east, the hoary-headed peaks of the Sangre de Cristo range loomed in the

distance, while to the west the sluggish Rio
Grande cut snakelike curves through the
valley floor, like a writhing prisoner caught
between the foothills of the Sangre de Cristos
and the San Juan Mountains.

The young boy lay on his back on a
low hill east of his pueblo. He was a member
of the Picuris village, whose ancestors had
lived in the area beside the tiny Rio Puebla
for more than 700 years—a culture that was
now being systematically eroded, diluted.
Unemployment, alcohol, and other related
problems had caused many of the young men
to leave the Pueblo village in search of new
opportunities. Now the women of the village
spent most of their time making pottery for
sale to tourists, and the few men who
remained in the village were those too old to
leave.

The boy thought little of such things
from his vantage point on the hill. He lay

with his legs somewhat spread, his toes pointing northwest and northeast, two of the ritual directions fixed by the points on the horizon where the sun rises and sets on the longest day of the year. To his left and right were the southeast and southwest, where the sun rises and sets on the shortest day. Above him was the fifth sacred direction, the zenith, home of the white-headed eagle who lived in the sky and, according to his grandfather, sometimes carried away young Pueblo boys to its sky-home. Below him was the sixth direction, the nadir, the home of the night birds and darkness.

Suddenly Tonita lifted his head and turned his ears to catch a strange sound. Beyond the hubbub of the village sounds, a new and different one was emerging. He listened still harder. To the east, he could hear the plaintive *pey-cos, pey-cos* calls of sagebrush-softness, the cottony-topped quail

(97)

that often ran through the sagebrush ahead of him, and still farther away a bird-of-the-morning-light uttered a raucous *mag-pie,* as if also asking what might be approaching.

The sound increased like an invisible army of clarion trumpets, whose calls drifted downward from the scattered clouds. As the boy sat up to shade his eyes from the lowering sun, he saw an undulating line in the northern sky, gently swaying like an inky specter, but slowly approaching. Now standing, Tonita watched quietly as the line soon separated into individual spots, which then enlarged to become birds, each with a gray body and wings, long trailing legs, and an equally long neck and head. Thus, each one formed a cross-shaped pattern in the sky that reminded him of the great cross-shaped constellation flying through the Milky Way. With powerful wingstrokes, the birds passed overhead, scarcely noticing the boy below, but intent on the distant point where the Rio Grande disappeared on the southern horizon.

Standing silently until the birds were out of sight and nearly out of hearing range, Tonita wondered if he had seen a vision or some new kind of sky-bird that lives with the eagle and never alights on earth. Then, throwing off his daze, he ran quickly down the hillside and back toward his pueblo as rapidly as he could. Nimbly jumping over the lower sagebrush shrubs, he skirted the ruins of the excavated kiva in the oldest section of the village, and found his grandfather dozing in the shade of their small frame dwelling.

"Grandfather, Grandfather, I have just seen a marvelous sight! Please wake up and tell me what I have seen! I saw almost 20 birds the size of a mountain turkey, but gray like the color of *kaipia-o-one,* the dove, and able to fly as easily as an eagle! And when they called, it sounded like a hundred bellowing sheep. They flew in a long line from the north to the south, and finally disappeared from my view."

The old man listened patiently and smiled. "Yes, my son, those are the crane people, who once lived here in great numbers, when the Rio Grande ran full. Their numbers were as many as the sagebrush on the hillside, and when they flew overhead they nearly darkened the sky. But in my lifetime the river has been nearly sucked dry by the white man's irrigation, and the cranes have been so shot at by hunters that they no longer come to our valley."

"Tell me more about them," Tonita said eagerly. "Tell me what they eat and where they live."

The old man thought a moment, and

slowly answered, "According to the legends
of our tribe, there was once a great flock of
cranes that lived in the clouds of the sky.
They drank the water from the clouds, and
they even built their nests in the clouds, and
they lived happily. Yet, they became tired of
this easy life, and the leader of the crane
flock one day gathered his tribe together and
said, 'Let us go down to earth, where the
waters have many fishes and frogs to eat.' So
the flock descended from the clouds and soon
found a small spring, but very shortly they
had drunk all of its water and eaten all of its
animals, and so they were forced to move on.
They then went to Taos, where the same
thing happened. Next they came to Picuris,
where they drank all the water of the Puebla
River. Finally, they were forced to leave
Picuris, and then they flew to the Rio
Grande, where they found abundant water
and food. There they made their camp, for

although the cranes did their best to drink
all the water and eat all the animals, there
was an abundance of these, and they could
not drink the river dry. Finally, the leader of
the flock said, 'This river must be very
strong, so here we will make our camp, here
we will build our nests and increase in
number.'"

Tonita squirmed impatiently. "But
Grandfather, I have never seen the cranes
along the Rio Grande in the summer, so how
can they nest here?"

His grandfather replied, "That was a
very long time ago, when there were many
birds that are no longer here. That was when
the white-headed eagle watched over us, and
when animals could talk to humans. Then
too we had a crane society in our tribe,
which was one of our secret societies. Each
member of the crane society carried a special
wand, with a tuft of feathers on it, including

two large crane feathers. Nobody else was allowed to harm a crane or touch its feathers. Our crane society was a guardian clan, because the cranes are so watchful and wary. Members of the crane society would guard the village to keep intruders out during secret ceremonies, and would also help to safeguard the passage of our dead to the other world."

"But what happened to the crane society?" Tonita was unaware that his people had ever had such secret rites. Although his village still annually performed corn dances in August, these were simply tourist-attracting performances, the last vestiges of the corn rituals that once were so important to the Pueblo peoples.

"Like most of our traditions, the crane society and our other secret societies have disappeared," the old man said softly. "Our lands have been taken by the white man, and our people have been poisoned by his ways. And so, our village is dying, just as the river is dying. Soon even you will grow up and leave the village, to live in Sante Fe or Taos, and you too will soon forget all the stories I have told you about our old ways."

Tonita was sadly quiet for a time, as he thought about leaving his village and family, and of no longer running happily in the desert sage, chasing lizards and following the tracks of animals in the sandy soil. And he wondered too about where the cranes had flown, and if they would find happiness in their new home.

Meantime, the cranes had already passed the Tewas villages of San Juan and San Ildefonso, and had continued southward past Santa Fe. The flock now numbered only 18 birds. These included six pairs without offspring, one adult female widowed a short time previously when its mate was killed by hunters in Colorado, a three-year-old unpaired male, a two-year-old subadult, and a pair leading a youngster hatched the previous summer in Alaska.

The juvenile usually flew between his two parents, trying desperately to keep up with them at times when they were flying into headwinds, and occasionally uttering a plaintive call when it seemed that the day's flying would never end. This flight had begun in southern Colorado, where the birds had spent a few days roosting along the sandy flatlands of the Arkansas River. Normally, they would have continued directly south from there, but strong easterly winds had blown them far off course, and they had found themselves in strange territory, crossing a high mountain range. Then the adults caught sight of a silvery line in the far southwest, surely a river, and it was flowing in a vaguely southerly direction. Thus it was that the cranes crossed into the Rio Grande Valley, and found themselves following a migratory pathway quite different from their usual route to their traditional wintering grounds on the arid plains of southeastern New Mexico.

The Staked Plains

As they had flown above the village of Picuris, the young crane was too tired from its long flight over the mountain pass to have noticed the Indian boy standing and watching him. Instead, he tried to maintain the right distance between his parents, close enough to see their every move and signal, but not so close that there was a danger of their wings striking his, for that could easily result in a broken wing or a badly bruised wrist joint.

The flock had no real leader as such; at various times each of the adults was temporarily in the lead, but in general the group flew in a fluid diagonal formation, often calling excitedly to one another whenever they saw possible landing places or dangerous hazards before them. Once a golden eagle followed them for a time, closely watching every bird to see if it might be straggling and weaker than the others, less able to avoid a sudden diving attack from above. Finally, the eagle drifted away, and instead began to look for jackrabbits in the scrubby foothills.

The cranes were becoming extremely tired; all day they had been buffeted by strong sidewinds and they had had to also ascend the mountain range. As they reached an area of the river where pasturelands and meadows came down to the water's edge, they began to circle the river cautiously and gradually started to lose altitude. None of the adults had ever visited the region before, and so the flock circled for nearly 15 minutes as every bird scanned the ground below for hidden hunters, coyotes, dogs, or any other threat. Finally, the birds cupped their wings, dropped their long legs, spread their tails, and began to parachute downward toward the water's edge. As they landed in the shallow water, each bird took a long drink, and then immediately waded out into the shallows to preen and rest.

The young crane quickly joined his parents in these activities. Soon the entire flock started to move to shore, and to probe in the still-green pastureland near the water. Here sedges and herbs still grew with a summerlike luxuriance, and the birds began to pull them up and consume them, roots and all.

Finally satisfied, the flock again

moved back to the safety of the water, just as the sun's disk was beginning to disappear behind the western horizon. It had been a long day, and although the birds might well have continued south and thus reached the wintering grounds of the greater sandhill cranes, they knew that they were far off course, and would soon have to fly back to the east and try to locate their own familiar wintering areas.

While the cranes huddled together in the darkness, a great horned owl called softly in the distance, trying to locate its mate, which had gone hunting at sunset and had not yet returned. The hunting owl passed over the river a few times, but the cranes neither saw nor heard it moving overhead on its silent wings. Clearly, the owl was outmatched by even the youngest of the cranes. Thus it ghosted onward, and finally pounced on a kangaroo rat that had carelessly moved too far from its burrow.

As the night passed, temperatures quickly cooled. Before dawn, ice had formed in the shallow backwaters of the river and around the legs of the roosting birds, who had to pull themselves free of the ice's grip. Luckily, none of them was trapped by the ice, nor had the ice become thick enough to allow a coyote to walk upon it and reach the birds.

Soon pale shafts of light appeared in the eastern sky, flickering through the clouds like the white underwings of a magpie, and the distant mountains were transformed from their midnight blue to blood-red, melting to a sandy pink. The cranes lifted their heads from their shoulder feathers, shook their bills to free the bits of down that stuck to their nostrils like giant white snowflakes, and began to wade to shore to forage a short time before taking flight. Soon the life-giving sun

would
form rising
currents of warm
air that would lift
them out of the valley
and high into the air,
and a westerly wind would
let them drift eastward across
New Mexico to their ancestral winter
homeland. It would be a homecoming
offering both security and danger. For almost
two decades the cranes had been legal game
on their wintering grounds of New Mexico
and Texas, which had been the first states to
allow "experimental" killing of the birds.
After 20 years, perhaps as many as 200,000
cranes had thus been "harvested" throughout
the areas where shooting was legal, nearly as
many as the total number of cranes still in
existence.

The vast high plains of eastern New
Mexico and northwestern Texas once were
covered by a sea of low desert grasses that
blanketed the otherwise featureless landscape
with such uniformity that it is said that the
early Spanish explorers periodically drove
stakes into the ground to mark their trails
during their fruitless quests to find Quivera,
the Kingdom of Gold. Later, the Comanches
used this remote and almost waterless land as
a safe refuge from which they could launch
their deadly attacks on the Spanish-speaking
settlers or make forays into Old Mexico.
Finally, even their legendary leaders were
forced into submission, and it again became a
forgotten corner of the New Mexican
Territory, a land of frequent duststorms
whipped up whenever the fragile sod was
overturned or too heavily grazed, and of
waters so bitterly alkaline that no human or
animal could drink them.

To this unlikely sanctuary the cranes

return each winter. In a narrow U-shaped valley the shallow Pecos River trudges its way reluctantly southward, occasionally gathering strength from tributary streams that usually survive just long enough to dump their salty waters into its own. At other times, the creeks simply die in the desert landscape, or disappear in the ground, perhaps to reappear some distance away. Thus, the Lost River appears above ground northeast of the town of Roswell, and joins Bitter Creek before entering Bitter Lake. Slightly beyond, shallow areas of seepage provide marshes that form the wandering outlines of ancient channels of the Pecos River.

The cranes, now flying single file in the morning sunshine, traveled the last few miles toward the remembered sanctuary at Bitter Lake with a sense of increasing urgency and eagerness. They were finishing a journey that had spanned an entire continent, and that they and their ancestors had traced and retraced for nearly 10 million years. Their pattern for survival had long ago been fixed into an endless seasonal repetition of tundra, mountains, and plains, and of recurring birth, death, and rebirth. At Bitter Lake they would finally rest and restore their strength during the short winter days. And so they would become ready for yet another spring flight northward, when the arc of the sun again begins to rise and give warmth and light to the northern lands. The cranes would continue to wade the earth's waters, tread lightly on its plains, and embrace the sky for as long as mankind allows them to do so.

(116)